by Michelle Anderson

Copyright © by Harcourt, Inc.

All rights reserved. No part of this publication may be reproduced or transmitted in any form or by any means, electronic or mechanical, including photocopy, recording, or any information storage and retrieval system, without permission in writing from the publisher.

Requests for permission to make copies of any part of the work should be addressed to School Permissions and Copyrights, Harcourt, Inc., 6277 Sea Harbor Drive, Orlando, Florida 32887-6777. Fax: 407-345-2418.

HARCOURT and the Harcourt Logo are trademarks of Harcourt, Inc., registered in the United States of America and/or other jurisdictions.

Printed in the United States of America

ISBN 978-0-15-362491-9
ISBN 0-15-362491-4

1 2 3 4 5 6 7 8 9 10   175   16 15 14 13 12 11 10 09 08 07

Visit *The Learning Site!*
www.harcourtschool.com

Introduction

In the summer of 1978, the veterinarians at the San Diego Zoo had a mystery to solve. Three of their white polar bears had turned green! After some investigation, the veterinarians discovered that a type of blue-green algae was living inside the polar bears' hollow hair shafts. How did this happen?

The polar bears' natural habitat is the Arctic, where they have adapted to extremely cold temperatures. The climate of San Diego, California, is warm. The algae growth was a result of this change in the polar bears' habitat.

Fortunately, the polar bears were not harmed by the algae growth. But their green coats can demonstrate something about how plants and animals adapt to their habitats. Most humans live in environments in which conditions are not too extreme. Some plants and animals, however, live in extreme environments. Imagine living in a desert that gets less than 5 cm (2 in.) of rain during the entire year, or at the bottom of the ocean where there is little food and no light. Plants and animals have adapted so that they can survive in extreme environments in all parts of the world.

The Namib desert

Hot, Dry Deserts

What do you think of when you hear the word *desert*? Words such as hot and dry probably come to mind. Almost every continent has a desert. You may have learned about the Sahara in northern Africa, the Gobi in Asia, or the Mohave in the southwestern United States. Some deserts are even located on coasts, such as the Atacama, which is along the Pacific Ocean in Chile, South America. Another coastal desert is the Namib, along the Atlantic Ocean in Namibia, Africa. These coastal deserts are two of the driest deserts in the world. Deserts can be hot or cold, but all deserts receive little precipitation. The plants and animals living in deserts must have special adaptations to survive the harsh environment and shortage of water.

The Namib desert gets an average of two inches of rain per year. Some areas of the Namib desert receive only 0.5 cm (0.2 in.) of rain per year! This desert is thought to be the oldest desert on Earth. Many plants in the Namib desert grow nowhere else on Earth. The Namib desert is one of the few fog deserts. Warm air from the desert blows off the land and cools over the ocean. When the air currents shift, a fog drifts over land, bringing cooler air with it. Many of the desert's inhabitants depend on this fog to survive.

Can you imagine collecting drinking water on your skin? The darkling beetle is one of the animals of the Namib desert that uses this technique. As the fog rolls across the Namib desert, this beetle goes to the top of sand dunes to catch drops of condensation from the fog. The condensation collects on the beetle's body and the water trickles into the insect's mouth for a refreshing drink! The fog is a source of water for other animals of the Namib desert as well. Other types of beetles dig small holes in the dunes and drink the droplets of condensation that form in them. The web-footed gecko allows water droplets from the fog to collect on its body. The reptile then uses its long tongue to collect the droplets.

Another adaptation that animals use to survive in hot deserts is avoiding the heat. Many of the smaller animals that live in the Namib desert burrow into the sand during the day to keep cool. Have you ever been to the beach on a hot day when the sand burns your feet? If you dig in the sand, though, you will find that the sand below the surface is cool. Reptiles, including the fringe-toed lizard and side-winding adder (a kind of snake), burrow under the ground to keep cool during the hottest part of the day. The side-winding adder buries itself in the sand with nothing but its eyes and the tip of its tail showing as it waits for its next meal to pass by.

Another Namib animal that buries itself in the sand is the small, blind, and earless golden mole. This mammal spends the daytime hours burrowed deep in the sand. At night it "swims" just under the surface of the sand searching for prey. The golden mole eats insects, such as termites, and legless lizards. To capture prey and avoid predators, the golden mole uses its keen sense of smell and its ability to feel even the tiniest vibration through the sand.

The Namib desert is also home to a wide variety of larger animals such as gemsboks, ostriches, black-backed jackals, and brown hyenas, as well as fur seals that live along the coast! Many desert mammals are nocturnal, but the gemsbok, a type of antelope, often travels far into the hottest portions of the desert during the day.

A Gemsbok

To deal with the intense heat of the day, gemsboks have a specially adapted circulatory system that allows the blood to be cooled before it reaches the brain. The gemsbok can also survive without drinking water for long periods of time. It uses the water it gets from the plants it eats to survive.

The plants of the Namib desert include grasses and some small shrubs. Many types of succulents, plants that store water in their leaves, as well as wild melons, live in the desert. Some animals in the Namib desert thrive on succulents and wild melons. The animals eat these plants because they have a high water content. The plants grow throughout the desert, sometimes even between the dunes.

Two examples of specially adapted plants in the Namib desert are *Welwitschia,* which is a type of gymnosperm, and *Lithops,* which is a succulent. *Welwitschia* grows only about 30 cm (12 in.) tall; however, it spreads out almost 3.5 m (10 ft) across the desert floor. This plant produces only two leaves. The leaves continue to grow throughout the plant's lifetime, which is estimated to be several thousand years. As the years go by, this plant's leaves become dry and torn, making the leaves look like ribbons. Like many of the animals of the Namib desert, *Welwitschia* depends on the condensation from the fog for the majority of its water supply.

Lithops, also known as living stones or flowering stones, are succulents. Like *Welwitschia,* they also have only two leaves. *Lithops* have adapted to have pigments in their outer and inner layers that camouflage them to look like stones. You can barely pick these plants out from the pebbly soil in which they live. This adaptation protects the plant from being eaten by animals looking for succulent, water-filled leaves.

Dark, Deep Sea

Until recently, we haven't known much about life on the very bottom of the ocean. The lowest areas of the ocean, starting at 1000 m (3,300 ft) from the surface, are sometimes referred to as the deep sea. Some obstacles have kept scientists from exploring the deep sea. One obstacle is the distance from the water's surface to the bottom. The ocean floor can be as far as 11 km (7 mi) below the surface. This distance is much greater than the height of the tallest mountain on Earth. Another obstacle is the high pressure of the water. Submarines and other equipment have to be specially designed so that they are not crushed by the pressure of the water as they travel down into the deepest parts of the ocean. Humans also have difficulty adjusting to such high pressure. The deep sea is also completely dark—sunlight doesn't travel all the way down to 1000 m (3,300 ft) and below—and very cold, 1.4°C (35°F).

Certain organisms have adapted to the extreme conditions of the deep sea, although estimates of how many organisms are living there vary greatly. Some marine biologists have estimated that as many as 10 million different species live in the deep sea, but other biologists think this number is too high. However, nearly everyone would agree that the deep sea has an unexpectedly diverse habitat. The organisms that have been identified there can be divided into two groups: benthic organisms and pelagic organisms. Benthic organisms live on the bottom of the ocean. Pelagic organisms can swim and float.

Food is scarce in the deep sea. For one thing, no plants live there. Because sunlight does not reach those depths, photosynthesis cannot occur and so plants cannot survive. Because they have adapted to the high pressure of the deep sea, the pelagic organisms cannot swim to the surface, where there is more food.

For these reasons, deep-sea animals have developed adaptations that conserve energy and make it possible for them to find food. Typical deep-sea fish are very small and move slowly. They have poorly developed muscles and skeletons. They either have very small eyes or are blind. Some have large mouths and expandable stomachs. How can these organisms find food?

The large-mouthed fish can just wait for any prey or organic matter to float by. No matter how large the object is, their expandable stomachs allow them to gobble it up. Some deep-sea fish use bioluminescence, which is light that some organisms give off. For example, the anglerfish has a clever way of using bioluminescence to lure its prey. One of the spines on the anglerfish's fins is shaped much like a fishing pole. The "pole" dangles in front of the anglerfish's mouth. At the end of this pole is tissue in which bioluminescent bacteria live. The anglerfish's prey sees the light and, thinking that it is something tasty to eat, moves closer. It's the anglerfish, however, that gets the snack.

In other parts of the ocean, above the deep-sea levels, fish have beautiful, bright colors. These fish use their varied colors as camouflage to keep them safe from predators. In contrast, deep-sea fish are often either all black, red, gray or off-white, or have no pigment at all. Coloration, which often is used as camouflage, is unimportant in the deep sea's complete darkness.

More is known about the benthic organisms of the deep sea than about the pelagic organisms. Benthic organisms are easier for researchers to locate and study. Many of the deep sea benthic animals are related to the animals we see in shallower water. The sea cucumber, brittle star, and squid have all adapted to live on the deep ocean floor.

Some of these deep-sea animals undergo a phenomenon called gigantism, in which they are huge in comparison to their shallow water relatives. One example is a shrimplike crustacean, *Gnathophausia*, which can grow to almost 13 cm (5 in.) in the deep sea. This is up to six times larger than these organisms grow in shallower waters. The sea spider is another example of gigantism. The sea spider is small in shallow waters, but species in the deep sea can be as large as 76 cm (30 in.) across!

A Deep-Sea Dweller—the Anglerfish

Warm Oases in the Deep Sea

Some locations of the deep sea are not cold at all. Some are very hot. These areas are underwater hot springs called hydrothermal vents. These vents are mainly located along the oceanic ridges, where continental plates are slowly moving apart. As the continents move away from one another, magma from Earth's interior flows out to form new ocean floor. The new ocean floor has lots of cracks and crevices in it. Cold ocean water seeps deep into Earth's crust. Magma heats the water to temperatures as high as 350°C (660°F). The hot water also dissolves metals such as iron and zinc, as well as sulfides from Earth's crust. The hot water spews out of the sea floor, often appearing black because of the dissolved metals. These vents are nicknamed "black smokers." The minerals in the water are often deposited around the vents in formations called chimneys.

A Hydrothermal Vent Ecosystem

These vents are like an oasis in the cold, dark deep sea. The warm, sulfur-rich water allows for an abundance of life that is not possible elsewhere at such depths. Thermophilic, or heat-loving, bacteria live in and around the vents and thrive in the super-heated water. The presence of these bacteria allows other life to exist around the hydrothermal vents. The bacteria form the base of the food chain. Photosynthetic organisms are at the base of most food chains. Recall, however, that there is no light at the bottom of the ocean. The bacteria in hydrothermal vents use the hydrogen sulfide in the water to undergo chemosynthesis. Chemosynthesis is a process in which organisms use chemical energy from energy-rich compounds to produce sugars. Hydrogen sulfide is toxic to most organisms. Some bacteria, however, thrive in its presence. Besides forming the base of the food chain, through chemosynthesis these bacteria also remove the toxic hydrogen sulfide from the water. This allows a variety of unique organisms including mussels, clams, worms, and fish to live around hydrothermal vents.

Thermophilic bacteria also form symbiotic relationships with other organisms in the hydrothermal vent community. One such relationship exists between the bacteria and a species of tubeworm. The bacteria live inside the tubeworm and produce nutrients for the worm through chemosynthesis. This worm has neither a mouth nor a digestive tract, and it relies on the bacteria for all its nutrients. In return, the worm provides the bacteria with the materials needed for chemosynthesis. The worm's hemoglobin is specially adapted to capture the hydrogen sulfide from the water and transport it to the bacteria. This protects the worm from any toxic effects of the hydrogen sulfide and provides the bacteria with much-needed material for chemosynthesis.

Clams, scallops, and mussels also have symbiotic relationships with these bacteria. Other animals, such as vent shrimp, scour the ocean floor around the vents and the chimneys for bacteria to eat. The food chain around these vents also includes fish and crabs. Other pelagic animals in the deep sea have been seen in vent communities for brief periods of time, foraging for food.

Icy, Arctic Waters

Have you ever gone swimming in ice water? Probably not. Humans cannot survive in icy cold water for more than a few minutes. However, a variety of animals inhabit just such an environment. Animals such as whales, seals, sea birds, and fish that make the chilly, arctic waters around the North Pole their home have special adaptations that allow them to live in such an extreme environment.

Bowhead Whale

One way in which the mammals living in the frigid waters of the Arctic stay warm is through insulation from blubber. Blubber is a very thick layer of fat that lies just under the skin of marine mammals such as whales, walruses, and seals. This layer of fat acts as insulation from the cold. It keeps the core of the body, along with all the important internal organs, protected from excessive heat loss. Blubber can be anywhere from 5 cm (2 in.) thick in some types of seals to 60 cm (24 in.) thick in the bowhead whale!

Mammals living in arctic waters also have an adaptation to their circulatory systems that keeps the core of their body warm. The limbs of a marine mammal usually have less blubber protecting them, so they can remain flexible. To prevent frostbite, blood vessels in the limbs contract, allowing only enough blood to the limbs to keep them from suffering frostbite. The blood vessels of these animals also have veins and arteries running close to each other so that the warmer arterial blood can warm the cooler blood of the veins as it returns from the limbs. This helps keep the core body temperature from dropping.

Fur is another insulator against the cold. Some marine mammals in the Arctic, including the sea otter and the northern fur seal, have such thick fur that water never reaches their skin. This adaptation serves the same purpose as blubber; it reduces heat loss and keeps the animal warm. The northern fur seal has as many as 54,000 hairs per square centimeter (0.16 in.^{2})!

Mammals are not the only animals living in arctic waters. There are numerous types of sea birds that live in the Arctic and frequent its icy waters. The best adaptation these birds have for staying warm is down. A soft, insulating layer of down is between their skin and coarse feathers. Down traps air close to the bird's body, which helps keep the bird warm.

Some arctic fish have a unique adaptation to life in freezing cold water. They have special proteins in their blood that act like antifreeze. Fish cannot internally regulate their body temperature, so they are in danger of freezing when the water temperature gets too low.

The proteins stop the formation of ice crystals in the fish's blood. This keeps the blood from freezing and killing the fish. Because of the proteins, the Alaska blackfish can survive in water temperatures as low as –20°C (–4°F)!

Tundra

Animals have adapted not only to life in the icy arctic waters but also to life in the tundra. The tundra is the environment north of the tree line. If you traveled north through North America, Europe, or Asia, you would eventually find that forested land gives way to a more barren environment with low-growing plants, or tundra. The average yearly temperature in the tundra is below freezing. During July, temperatures rise to about 5°C (41°F). The average winter temperature is −30°C (−22°F). The growing season for plants in the tundra is less than two months.

The tundra is home not only to a variety of animals, but also to a variety of plants and lichens. Plants living in the tundra include two types of dwarf trees and many perennial plants. Perennial plants are plants that live year after year. Plants in the tundra have adapted to being perennials because the growing season each year is so short that it takes many years for them to fully grow. Arctic plants also grow very low to the ground. This protects them from the wind and increases their chances for being protected by snow in the winter. Snow is an excellent insulator, and both plants and animals benefit from spending the winter beneath the snow.

Animals use the insulating quality of fur to stay warm on the tundra. For example, the muskox has two layers of fur. The overlayer is a thick fur that is waterproof. The underlayer of fur traps a layer of air along the muskox's body. This air is heated to the animal's body temperature and aids in keeping it warm. Moose have club-shaped hairs that are thinner toward the skin than at the tip. This adaptation acts similarly to the down of the muskox; it traps a layer of air near the moose's skin, which helps keep it warm.

The polar bear's fur and thick layer of fat help keep it warm. The hairs that make up the fur of the polar bear are hollow and transparent. This allows ultraviolet light from the sun to reach the polar bear's black skin. The black skin absorbs the sun's heat. The combination of the absorption of solar heat and the insulation provided by fur keeps the polar bear warm, even in extremely cold temperatures.

Like the sea birds that spend time hunting in the arctic waters, the birds of the tundra use feathers, including down, to keep warm. Arctic birds include owls, many migratory species, and the ptarmigan. The ptarmigan is covered from head to foot in feathers. The feathers on its feet and toes help keep the bird warm and allow its feet to act as snowshoes while the bird walks on the snow looking for berries and leaves to eat. This bird uses a number of adaptations to keep warm and safe. In addition to feathers, the ptarmigan burrows in the snow at night to keep warm. Each night these birds fly up into the air and then dive into the snow! They do this so that predators will not see their tracks leading to their burrow. The ptarmigan leaves no trace behind and can sleep safely.

The ptarmigan has another adaptation for survival—camouflage. In the summer the birds are brown, but in the winter their feathers are snowy white. Other animals of the arctic use camouflage as well. The arctic ermine, the arctic hare, and the arctic fox all have white coats during the winter. Camouflage keeps these animals safe from predators.

A Ptarmigan

Some animals avoid winter by sleeping through it in their dens. These animals are not truly hibernating. They are dormant, but their body temperatures remain almost normal and they can be awakened. When animals are in true hibernation, they cannot be awakened. One mammal in the arctic that truly hibernates is the ground squirrel. These squirrels spend the summer eating as much food as possible. In the winter, they enter their dens to hibernate. Their breathing, heartbeat, and body temperature all drop dramatically.

Many arctic animals avoid the harsh winters altogether. Most bird species in the arctic are migratory, and they leave before winter sets in. The majority of the caribou herds migrate south also, along with many of their predators, such as wolves. These animals will return to the Arctic in the spring, when the tundra is more welcoming.

Hypersaline Lakes

You already know that the ocean contains salt water, but did you know that some lakes have water that is even saltier than ocean water? These lakes are called salt lakes, or hypersaline lakes. The ocean has a salinity of about 3.5 percent, but salt lakes have a salinity of 10 percent or higher. Most salt lakes form when rivers flow into lakes that have no water source flowing out of them. The rivers flowing to the lakes pick up minerals, including salt, from the soil and rock that they flow over. These minerals dissolve in the river water and are dumped into the lakes. As evaporation occurs in the lakes, the concentration of salt increases and the lakes become hypersaline environments.

The Caspian Sea in northern Iran is the largest salt lake on Earth. The Great Salt Lake in Utah and the Dead Sea, located between Jordan and Israel, are both hypersaline environments. The Great Salt Lake's salinity fluctuates but is usually near 20 percent. The Dead Sea is the saltiest of all, with a salinity of over 30 percent!

Very few plants or animals can tolerate the extreme salinity of the Great Salt Lake or the Dead Sea. The imbalance of salt molecules and water molecules between an organism's internal environment and the external environment causes water to be pulled from an organism's cells. This causes the cells to dry out and die, which is why most organisms cannot survive in hypersaline environments. However, some organisms have adapted to these conditions.

Halobacteria are halophytes, or salt-loving organisms. Halobacteria live in hypersaline environments, including the Dead Sea. These bacteria not only survive in salt lakes, but thrive in them. In fact, many species of halobacteria cannot live in environments with less than 15 percent salinity. Halobacteria can move salt into their cells to create a higher salinity than the surrounding water. By accumulating salt in their cells, they can maintain a balance between the internal and external environment of their cells. Water is not pulled from their cells, and they do not dry out in the hypersaline environment. Besides several species of halobacteria, one species of alga is found in the Dead Sea. When the alga blooms, the lake water turns reddish in color.

Because the Great Salt Lake has a lower percentage of salt in its water than the Dead Sea, the Great Salt Lake has a wider variety of organisms in its environment. One of the most common species of the Great Salt Lake is the brine shrimp, *Artemia*.

The Dead Sea

These tiny brine shrimp feed off the halobacteria and algae that also live in the Great Salt Lake. In turn, large numbers of migratory birds stop at the Great Salt Lake to feed on the brine shrimp. An industry has even formed near the Great Salt Lake to package these shrimp as food for tropical fish that live in aquariums.

Human impact on saline lakes can be seen in more than just the industry that has grown near them. People have lived near the Dead Sea for thousands of years. Over time, water from the many rivers that feed into the Dead Sea, including the Jordan and Arnon Rivers, has been diverted for irrigation. This has reduced the flow of water into the Dead Sea and has increased its salinity drastically. Many people are concerned about how these changes will affect the environment in and around the Dead Sea. They fear that the Dead Sea is actually a dying environment.

The Aral Sea is a saline lake located in Central Asia. Two rivers end at the Aral Sea, but much of their water has been diverted for crop irrigation. Before the 1960s, this sea was the fourth-largest inland sea. It has since shrunk to less than half of its earlier size. Cities that were once thriving fishing ports are now located tens of kilometers away from the current shoreline. The land around the current Aral Sea is now basically a desert, and the ground is covered in so much salt that it looks like snow.

The Aral Sea